Contents

Words in *italic* in the text are explained in the glossary on page 30.

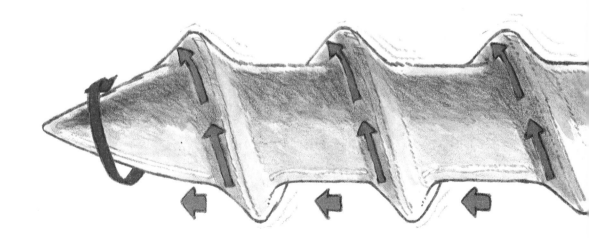

Getting to the top

Have you ever climbed a mountain? Or walked up a hill? If you have, you will know that the steeper the slope, the harder it is to walk up it. This has nothing to do with your legs, or how fit you are. It is a law of nature.

Raising an object (like you) by a certain distance (to the top of a hill) takes a fixed amount of work, or energy, no matter how you do it. But you can make it easier, by changing the way you do it.

This is because work is a combination of two things: effort (or force) and distance.

You can do the same amount of work, using the same amount of energy, by using either a lot of effort over a short distance, or a little effort over a long distance. The steepest route is shortest and hardest. A gentle slope is easier, but longer. Either way, you have done the same amount of work by the time you have reached the top.

On this steep mountainside, walkers have made a zig-zag path by walking across the slope, instead of straight up it. It is longer, but much easier.

SIMPLE SLOPES

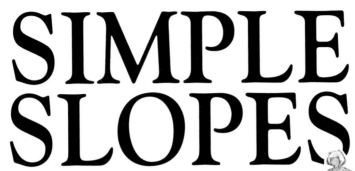

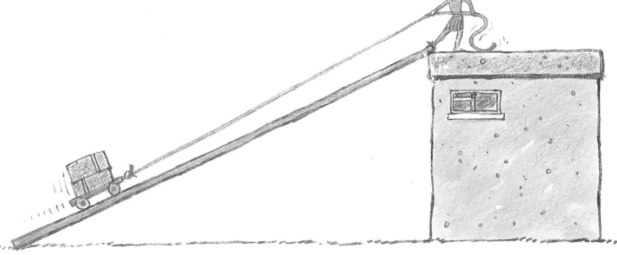

ANDREW DUNN

Illustrated by
ED CARR

Titles in this series
Heat
It's Electric
Lifting by Levers
The Power of Pressure
Simple Slopes
Wheels at Work

First published in 1991 by
Wayland (Publishers) Ltd
61 Western Road, Hove
East Sussex, BN3 1JD, England

© Copyright 1991 Wayland (Publishers) Ltd

Editor: Anna Girling
Design: Carr Associates Graphics, Brighton

British Library Cataloguing in Publication Data
Dunn, Andrew
   Simple slopes.–(How things work)
   I. Title   II. Series
   372.3

HARDBACK ISBN 0-7502-0217-3

PAPERBACK ISBN 0-7502-0955-0

Typeset by Dorchester Typesetting Group Ltd
Printed in Italy by G. Canale & C.S.p.A. Turin

What is work?

A short, steep slope is harder to climb . . .

. . . than a long, gentle one . . .

Less effort over long distance

WORK

More effort over short distance

. . . but the amount of work done is the same.

Lifting loads

The simplest way to lift an object is to haul it straight up. But this is not the easiest way, because it takes the most effort.

In the picture below, the *ramp* is three times as long as the height of the wall. So it takes only a third of the effort to lift the load to the same height. The person is using less effort over a longer distance.

The ramp is one of the oldest *machines* there is. The ancient Egyptians used ramps to build their *pyramids*. They had no *cranes*, so to raise the enormous blocks of stone, they used ramps. You can still see ramps on building sites today.

Nowadays we use the ramp – a simple slope – in all sorts of ways. There are ramps in everything from ploughs to zip fasteners, from keys and locks to screws, nuts and bolts.

6

How to build a pyramid

Wedges

A very useful form of slope is the wedge. It appears in many machines in different disguises. It can be a machine by itself – a door wedge is a good example.

When you use a door wedge, instead of moving an object up the slope, the slope moves to lift the door. As it is pushed under the door, it raises the door slightly with a lot of force. The door presses back down on the wedge with *equal* force, so that the wedge grips the floor firmly and holds the door open.

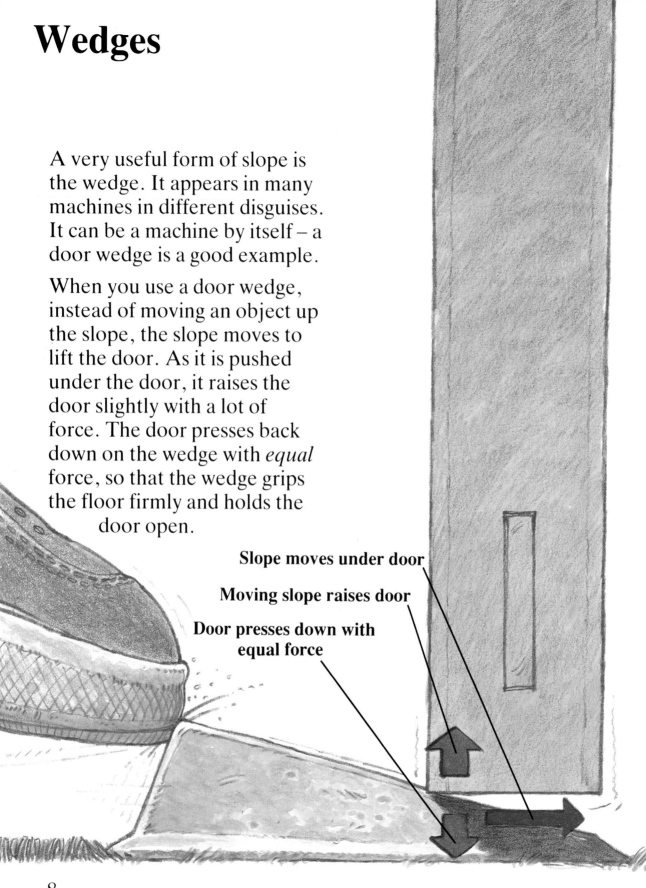

Slope moves under door

Moving slope raises door

Door presses down with equal force

Arches

Another example of a wedge is the stone used at
the top of an arch. It is called the keystone, and it
presses on both sides of the arch, forcing
the other stones together and giving
the arch great strength.

Next time you see
an arch like this, have a
close look. Is there a wedge-shaped
stone at the top?

Cutting wedges

An axe is a wedge with a sharp edge, attached to a handle.

A wedge is really made up of two slopes, back to back. When an axe strikes a log, the moving slope on each side pushes the wood away sideways, splitting it in two. The effort is *magnified* by the length of the slopes.

Most of the machines that we use for cutting – scissors, shears, can-openers, hedge-trimmers – involve wedges.

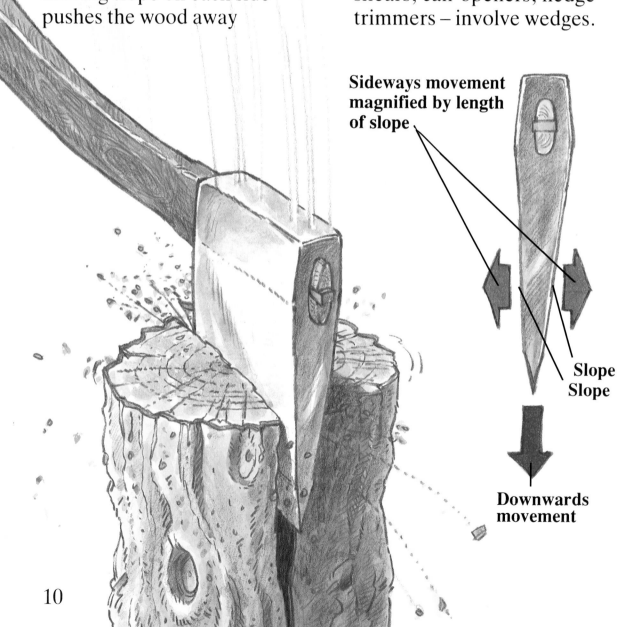

Sideways movement magnified by length of slope

Slope
Slope

Downwards movement

Scissors

Scissors are made of two wedges that cut through cloth or paper by slicing into it from opposite directions. As the *blades* meet, they act like one wedge, forcing the material to part sideways. Garden shears work in exactly the same way.

Hedge-trimmers

An electric hedge-trimmer uses two blades which slide backwards and forwards over each other. Each blade has many slots, each with a sharp wedge shape.

As the blades move, the slots open to allow a twig or leaf to enter the gap, and then close again, snipping the twig just like a pair of scissors.

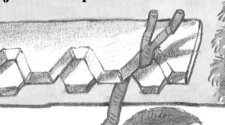

So using a hedge-trimmer is like using dozens of pairs of scissors very close together, all at the same time.

The plough

The ploughman's wedge

When farmers plough their fields, they cut into the top of the soil, lift it and turn it over. This breaks up the soil and mixes some air and old plants with it, so that it is ready for planting a new crop. A plough is a set of wedges, dragged through the soil by a tractor, a horse or an ox. It has been used by farmers for thousands of years. The only difference is that nowadays it is made of metal, instead of wood.

A plough is in fact made of three wedges. The first one cuts a slice down through the soil.

First wedge

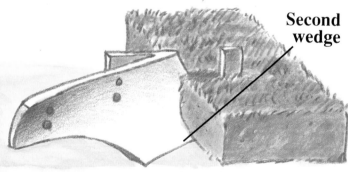

The second cuts sideways into the soil under the surface, separating a layer of earth.

Second wedge

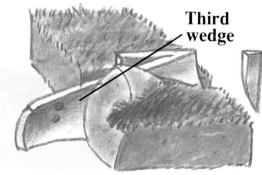

The third lifts this layer and turns it over. All the wedges produce big forces as they are dragged through the soil.

Third wedge

Modern ploughs work in exactly the same way as the ones used by farmers thousands of years ago. Can you see the wedges cutting through the soil?

The zip fastener

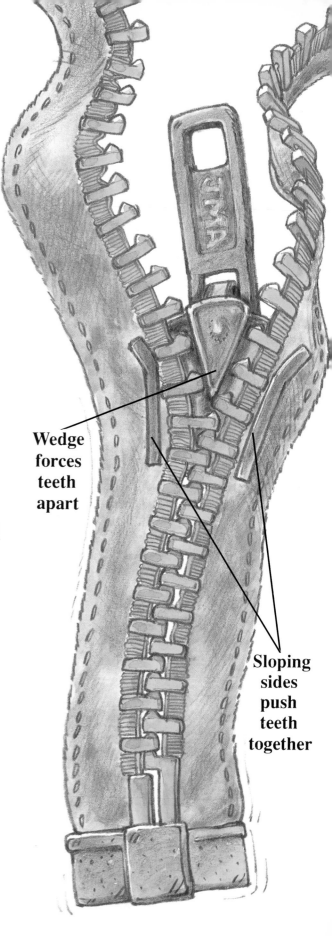

The zip fastener is a clever *device* which changes the small effort of pulling on it into a strong force that pushes tiny teeth together, or separates them. Each tooth slots into the one above it.

In the middle of the fastener slide is a wedge. If you have ever had a stuck zip, you will know that without the wedge in the slide it is very difficult to separate the teeth of the zip.

When you close the zip again, the two outer sides of the fastener also act as wedges, and push the teeth back into place.

Wedge forces teeth apart

Sloping sides push teeth together

The zip was invented in 1891 for doing up boots.

14

The wedge in nature

The wedge appears in nature as a very powerful tool.

If a crack appears in rock, water seeps into it when it rains.

During the winter, when it freezes, the water *expands* as it turns into ice. The sideways force is tremendous, enough to break the rock.

This is what causes roads to break up in winter, and makes *boulders* break away from cliffs.

The key in the lock

There are many machines that use the principle of slopes without using wedges. Have you ever looked carefully at a door key? Even if you have, it is not easy to see how it works, because you also need to look inside the lock!

Cylinder locks

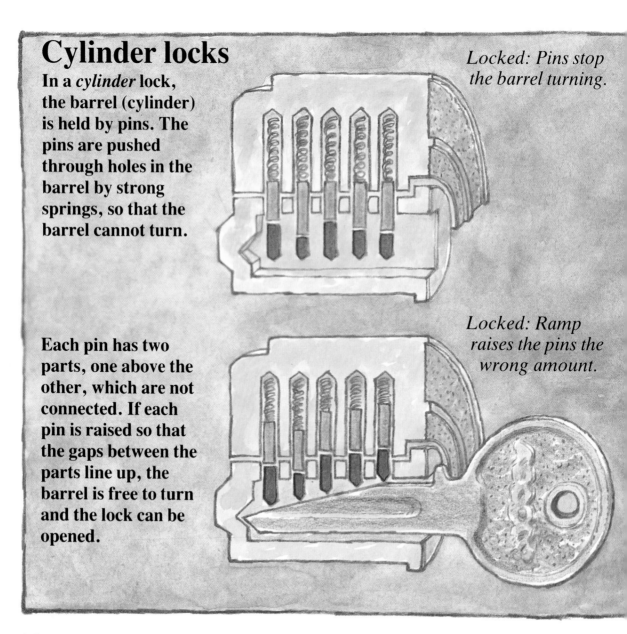

In a *cylinder* lock, the barrel (cylinder) is held by pins. The pins are pushed through holes in the barrel by strong springs, so that the barrel cannot turn.

Each pin has two parts, one above the other, which are not connected. If each pin is raised so that the gaps between the parts line up, the barrel is free to turn and the lock can be opened.

Locked: Pins stop the barrel turning.

Locked: Ramp raises the pins the wrong amount.

But the gap in each pin is at a different point, so each has to be raised by a different amount.

The answer is to make a key which has a different sloping ramp for each pin. As the key is slipped into the lock, each ramp pushes up the pins until the key is in the correct position to raise all the pins the right amount.

Of course, the key has to slip out of the lock as well, so in between each ramp is another, facing the opposite way.

When the correct key is in the lock, the barrel can turn and pull back the bolt to unlock the door.

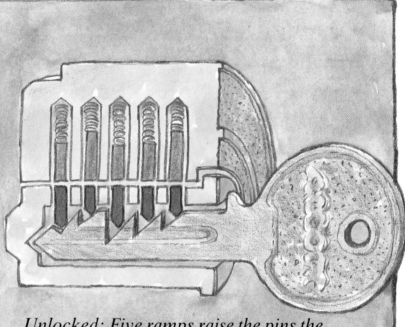

Unlocked: Five ramps raise the pins the correct amount but the key is stuck.

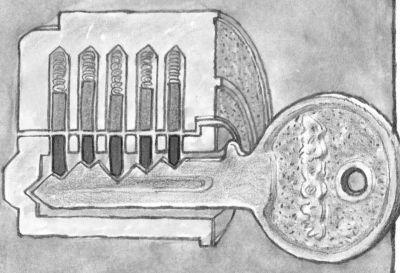

Unlocked: Pairs of ramps let the key slip out.

The screw

Take a look at a screw. Can you see the slope? A screw is made up of a long slope cut round and round a metal rod. The slope works in the same way as it does in a ramp. Because of the length of the slope, it produces a great force with little effort.

When you drive a screw into a block of wood, the screw itself travels only a short distance. But as it turns, the slope around it travels much further, magnifying the effort of turning and driving the screw powerfully forward. After a few turns, the screw is held firmly in place.

A helter-skelter is a slide which twists round – just like the slope of a screw does.

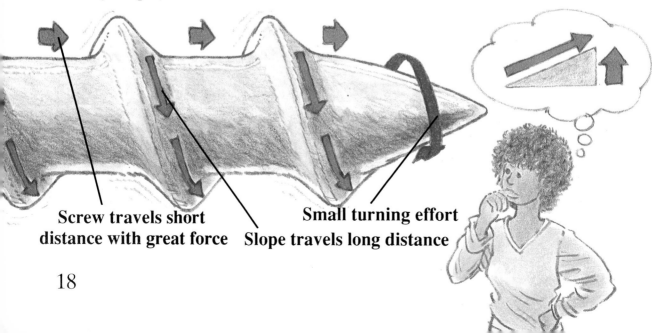

Screw travels short distance with great force

Small turning effort

Slope travels long distance

18

Nuts and bolts

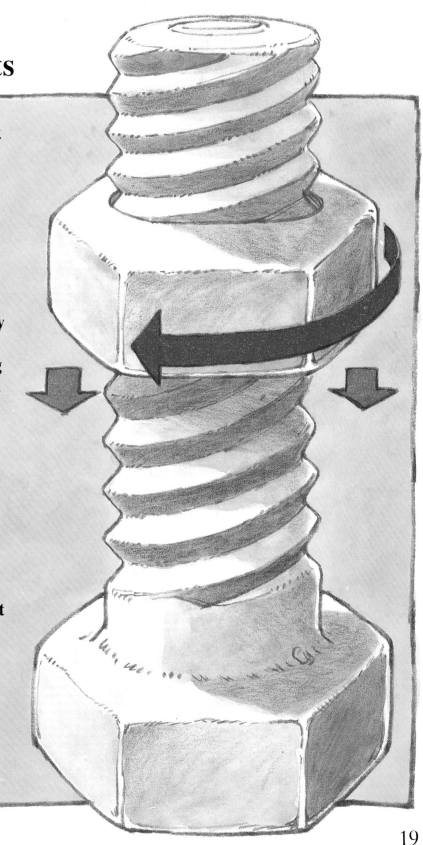

Nuts and bolts work in the same way as screws.

A nut has to turn round the bolt many times to travel a short distance along it. The force needed to turn it is very low, because the slope is very long and gentle.

But because the nut moves along the bolt very slowly, the force it produces is very great. This powerful grip is what makes nuts and bolts hold objects together so tightly.

Screws at work

Corkscrews

A corkscrew is a pointed *spiral* of metal and it works just like a wood screw. It can be screwed into the cork easily, and grips the cork well.

Jacks

A jack is used for lifting a car easily off the ground to change a wheel. The screw has to turn many times to lift the car a few centimetres. If the handle moves round fifty times as far as the car moves up, then it is fifty times easier to lift the car than it would be without the jack.

Looking at taps

Try this experiment. Turn on the kitchen tap and then try to stop the water, using your finger. Be careful, or you will soak yourself and the kitchen!

Do you see what enormous *pressure* the water has? Yet a tap can stop the water easily.

The effort of your hand is first magnified by the handle, which acts as a *lever*. Inside the tap there is a screw, which forces the *washer* down into the hole with great force.

The screw has a steeply sloping thread, so that it does not take many turns to close or open the tap.

Archimedes' and other screws

One of the earliest uses of the screw as a machine was by Archimedes. He was a very clever mathematician who lived in ancient Greece over two thousand years ago. He made a machine to lift water out of the *hold* of a ship.

Inside a watertight cylinder he put a spiral slope. As the handle was turned, the screw lifted the water up the spiral.

Archimedes' screw is still used in some parts of the world today, to help water crops.

These farmers are using a version of Archimedes' screw to raise water from a stream to their fields.

Augers

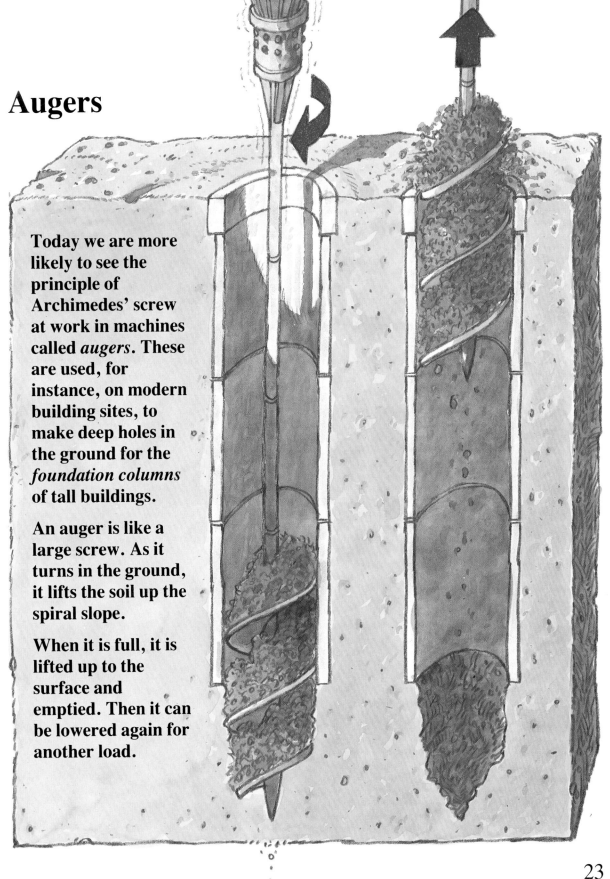

Today we are more likely to see the principle of Archimedes' screw at work in machines called *augers*. These are used, for instance, on modern building sites, to make deep holes in the ground for the *foundation columns* of tall buildings.

An auger is like a large screw. As it turns in the ground, it lifts the soil up the spiral slope.

When it is full, it is lifted up to the surface and emptied. Then it can be lowered again for another load.

Other augers

This auger is being used to drill holes for the foundations of a new building.

Mincers

You can find an auger in a butcher's shop, if he has an old-fashioned mincer.

The mincer handle turns both an auger and the cutting blades. The auger moves the meat forwards, forcing it past the blades which mince it, and out through the holes in the front.

Augers are used in tunnelling through soil and rocks. In a machine called a tunnel mole, cutting blades scrape away at the soil or rock face as they turn. Behind them, an auger moves the scrapings away from the cutting face.

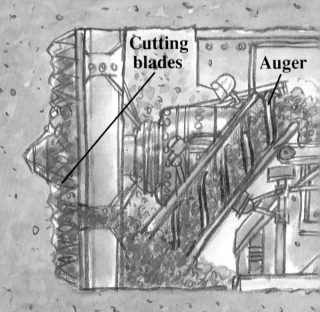

Cutting blades

Auger

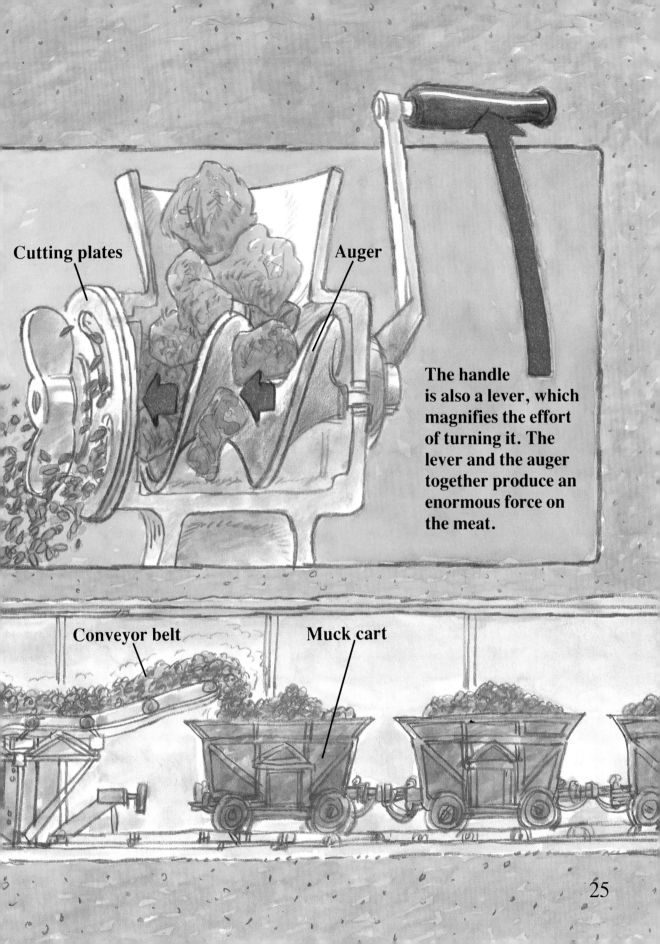

Cutting plates

Auger

The handle is also a lever, which magnifies the effort of turning it. The lever and the auger together produce an enormous force on the meat.

Conveyor belt

Muck cart

Drills

All drills are similar to augers, because they use the slope of the screw to carry material away. If the material stayed in the hole as it was drilled the *drill bit* would quickly jam solid and stick fast. So, as the drill cuts forward with its sharp point, the waste is carried away backwards along the screw of the drill.

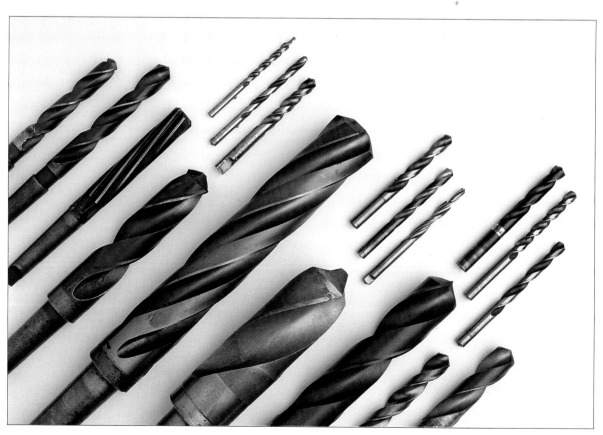

Drill bits come in many sizes. It is important to use the right one for the job.

In big drills, the grooves are deep and the slopes very gentle.

Have you ever seen metal being drilled? The waste that comes out has been bent around the drill, so it often has a corkscrew shape, like a pig's tail.

In fine drills, the grooves are very shallow and the slopes very steep.

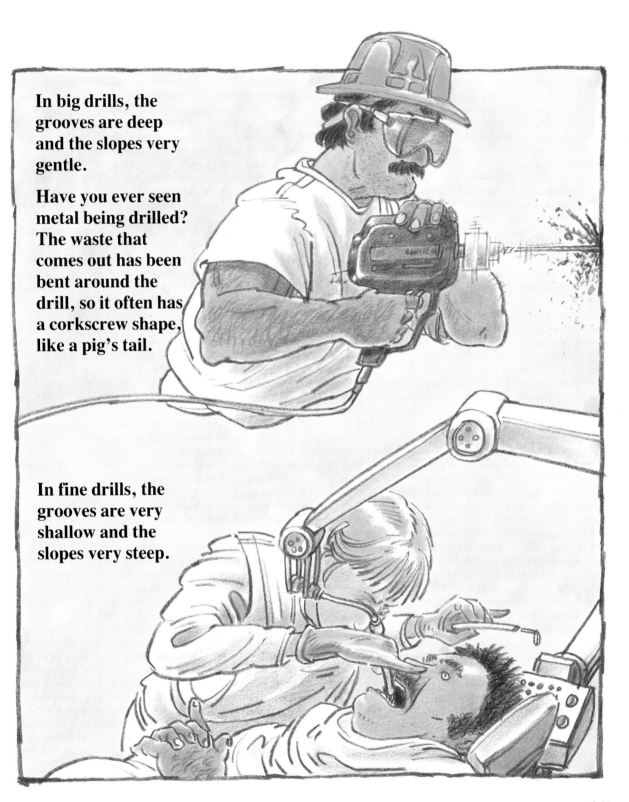

27

The simplicity of slopes

At first sight, the simple slope, or the inclined plane as engineers call it, looks too simple to be a useful machine. Yet even a wedge can be a very powerful tool. Many machines have extremely simple principles behind them. Even the most complicated machines, full of electronics, are often based on very simple ideas.

The bow of a ship is like a huge wedge that cuts easily through the water.

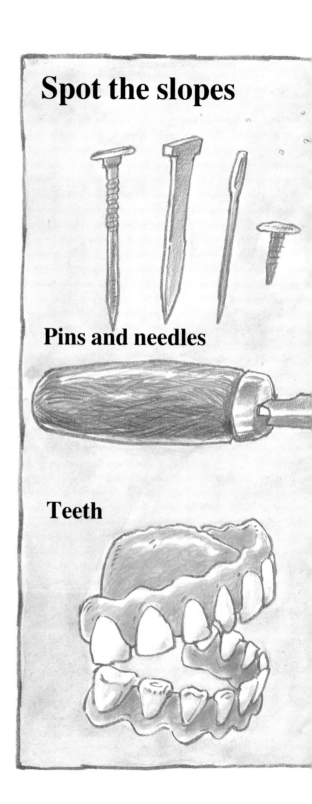

Spot the slopes

Pins and needles

Teeth

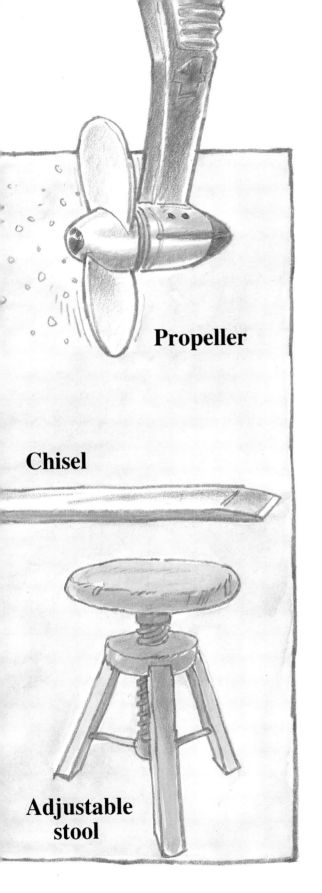

Propeller

Chisel

Adjustable stool

Propellers have sloping blades which 'throw' the air backwards, making the aircraft move forwards.

Next time you use a tool or a machine, or ride in one, or see one at work in the street or on television, see if you can find the simple idea that lies at the heart of it. You will be surprised how often you can!

Glossary

Augers   Pieces of machinery, shaped like screws, often used for drilling holes in the ground.

Blades   Sharp cutting edges, for example on knives, scissors or lawnmowers.

Boulder   A very large rock.

Cranes   Tall towers with long arms at the top, used on building sites.

Cylinder   A round tube. Cylinders can be solid or hollow.

Device   Another word for a small machine. It is often used to mean a clever invention.

Drill bit   The part of a drill that does the cutting.

Equal   Two things are equal when there is the same amount of each.

Expand   To grow bigger.

Foundation columns   Deep, round pillars of strong concrete sunk into the ground to provide a solid base for large buildings.

Hold   The part of a ship where the cargo is stored.

Lever   A bar which balances or turns on a point, called a fulcrum. By using a small force to move one end of the bar a long distance, you can produce a bigger force at the other end. Like a slope, a lever is a simple machine.

Machines   A machine is anything made by people to make work easier to do.

Magnified   Made to appear bigger or greater.

Pressure   The amount of force with which something presses on something else.

Pyramids   The large buildings that the ancient Egyptians made to hold the body of a dead king or queen. A pyramid has triangular sides which meet at a point at the top.

Ramp   A smooth man-made slope between places of different heights.

Spiral   A shape like a coil.

Washer   A flat, rubber ring used in a water tap. When the tap is closed the washer is pushed down tightly to stop any water dripping out.

Books to read

For younger readers:
How Machines Work by
Christopher Rawson (Usborne
Publishing, 1988)
How Things Work by Robin
Kerrod (Cherrytree Books, 1988)
Levers and Ramps by Ed Catherall
(Wayland, 1982)

For older readers:
Exploring Uses of Energy by Ed
Catherall (Wayland, 1990)
Machines by Mark Lambert and
Alistair Hamilton-MacLaren
(Wayland, 1991)
The Way Things Work by David
Macaulay (Dorling Kindersley,
1988)

Picture acknowledgements

The publishers would like to thank the following for providing the
photographs for this book: Cephas Picture Library 24 (Nigel Blythe), 29
(D. Burrows); Chapel Studios 18, 26; Frank Lane Picture Agency 13
(Maurice Walker); G.S.F. Picture Library 22; PHOTRI 28; Science Photo
Library 14 (John Heseltine); Tony Stone Worldwide 4.

Index